UFOs

UNEXPLAINED FLYING OBJECTS

BY PG MUCKENHAUPT

COPYRIGHT ALL RIGHTS RESERVED

BOOK 9 BY PG MUCKENHAUPT

OTHER BOOKS BY THE AUTHOR

TREASURE PIRATES SHIPWRECKS GOLD AND SILVER

MORE TREASURES SHIPWRECKS GHOST TOWNS AND OLD FORTS

THE ULTIMATE MARTIAL ARTS CHUNG DO KWON AND KENDO

ALTZHEIMERS IN THE FAMILY

GHOST STORIES IN AMERICA

APOCALYPSE 19

A STUDY OF THE STRONTIUM GEOCHEMISTRY IN THE FORT RILEY LIMESTONE

SCUBA SNORKEL SWIM

AVAILABLE ON EBAY AND AMAZON.COM

DEDICATED TO LANCE AND ELLEN

INTRODUCTION

Ufos or as I call them, Unexplained Flying Objects, have been observed throughout history, Particularly during the Renaissance era. Masters have included them in their works. They have also been observed in cave paintings, The Mahabarata, the Pentagon, WW2, Roswell notably, all over the world, Area 51 in Nevada, by the Air Force and Navy in particular. The Bible also mentions them. Thousands of pictures and videos have been taken, some of them are unexplained to this day.

Of particular notice is the Pentagon releasing once classified videos from actual Navy sightings.

The technology exhibited by these craft exceeds current capabilities of our species by many years, hundreds or perhaps thousands. Many serious efforts have been made to study this phenomenon by governments all over the world. The US has had several, including Project Blue Blook, Project Grudge, The Condor Project among others.

Table of Contents

Chapter 1 A Historical Perspective

Chapter 2 Sightings

Chapter 3 Photos

Bibliography and References

Chapter 1 A Historical Perspective

The latest divulge of information about UFOs came from of all places, the Pentagon. The AATIP, Advanced Aerospace Threat Identification Program released three videos taken by our Navy. This is no quack organization. Luis Elizondo, is a senior member of the Office of the Under Secretary of Defense for Intelligence (OUSDI).

Robert Bigelow, billionaire and owner of Bigelow Aerospace, stated that "Absolutely aliens exist".

Mr. Bigelow bought the Skinwalker Ranch, a famous locale for viewing unexplained phenomenon.

The NIDS, or National Institute for Discovery Science is also his company.

Now, let us discuss the Navy sightings. Commander David Fravor and the Nimitz Strike Carrier Group, had a sighting in November 2004. Marine Lieutenant Colonel Douglas Kurth investigated the radar observations. Kurth worked for Bigelow until 2013.

Bigelow received millions of dollars of funding from the US. A 494 page report resulted in data summaries, tables, charts, descriptions including biological field effects, physical characteristics, methods of detection, witnesses, photographs and synopsis all related to UFOs. Dozens of names were referenced including, Jacques Valles, Puthoff, Davis and Colm Kelleher. This report discusses the biological studies of extraterrestrial intelligences, intellectual debates, physiological effects observed from the UFOs, disclosure of UFOs to the general public, classified Blue Book project files that remain unexplained using current technological knowledge, and aerospace violations by these craft over restricted air space such as military facilities including nuclear facilities, which appear to be of much interest.

Monthly reports were sent to the Pentagon. As part of this study, by Bigelow.

An example of the project lists as a sighting in which an orange colored object, shaped like a "football" with enormous size, estimated to be that approaching a "foot ball field" was observed at the K-7 site. The SAT team was authorized to proceed to the K-7 site but they were intimidated by the object. The object then rose to 1000 feet above the missile site. NORAD picked up the object on radar. Two F-106 fighters were sent to intercept from Great Falls, MT. The object then ascended to 200,000 feet, where it was out of radar range. The missile at the site actually had different target numbers! The individuals of the SAT team were treated for psychological trauma.

Returning to the Navy sightings, the object was observed at approximately 50 feet above the Pacific Ocean. When a jet was scrambled to intercept, it moved VERTICALLY to 50000 Feet in approximately THREE SECONDS! This is amazing. The technology exhibited by this object is extraordinary. It approaches velocity approaching the speed of light! Way beyond our current abilities, especially inside the atmosphere.

Our current propulsion systems can only fractionally approach this, and the materials we currently use to construct our ships could not withstand the stress exerted by this rapid acceleration. Additionally, the human body cannot handle that sort of G force, not EVEN CLOSE!

Black budget intelligence programs are rarely if ever disclosed to the general public. There is a core of secrecy surrounding these activities. Not even information released through the Freedom of Information Act is complete.

"Field Effects on Biological Tissue" is a technical paper refers to the other title known as "Clinical Medical Acute and Subacute Field Effects on Human Dermal and Neurological Tissues". This paper, interestingly, studies the clinical medical signs and symptoms and biophysics of injury known and expected from near field NIEMR microwave, thermal, and unintended exposure to anomalous systems, aka UFOs!!!!

This paper was written by a well respected and highly educated neuroscientist, Christopher Green. The Pentagon spokesperson for AAWSAP was Gough; and the researcher involved was Roger Glassal. Research analysis was performed by Keith Basterfield. AAWSAP initially was given 10 Million US for the commencement of studies. Another 12 Million US was given in 2010.

The Canadian UFO records number about 750. They are held by the Canadian Department of National Defense and the Canadian Research Council. UFO records are held throughout the world, particularly the UK, Sweden, Denmark, Greece and Australia.

Dr. J. Allen Hynek founded the Center for UFO studies (CUFO). He was a professor at Northwestern University and a professional Astronomer. He previously worked at Harvard University with the Smithsonian Observatory. Approximately 701 out of the 12,000 sightings remain totally unexplained. This is from 1947 to 1969. The MUFON, or Mutual UFO Network is located in Bellvue Colorado. In the Soviet Union, many sightings were reported relating to the testing of secret military rockets. China also has a record of these mysterious objects. The UFOs typically change color, shape, speed and direction uncharacteristic of our technology. The objects often do not exhibit any sound observable by humans; however it is believed the propulsion is electrical in nature. The noise is beyond the range of human hearing, but some animals such as birds, cats and dogs have hearing significantly more sensitive to that of our hearing range.

Chapter 2

SIGHTINGS

There have been thousands of UFO sightings. We will focus on the most studied sightings here.

Tuesday, June 24, 1947. Kenneth Arnold, clocks nine objects moving in excess of 1200 mpg westward over Washington State. The objects were the size of a "DC-4". No known crafts were able to travel at that speed at that time. The objects were shaped like "pie tins" and were of metallic composition. The altitude was 10,000 feet. He stated that the objects performed maneuvers beyond technology at that time in history. Ten miles south of Mineral WA, Sydney Gallagher, also observed the objects flying in a formation. This information was submitted to the USAF. They responded that they had no aircraft in this area at that time. The observers noted that the craft had NO WINGS and exhibited no obvious methods of propulsion, i.e. NO gaseous exhaust associated with planes.

In July 1947, a craft crashed in Roswell NM. The craft crashed on a ranch owned by Mac Brazel. The object left behind debris, consisting of unusual metal, pieces of which were retrieved by the military. Major Jesse Marcel and Captain Sheridan Cavitt investigated. They were under the orders of Colonel Blanchard. The debris were sent to a military base in Ft Worth TX. The area was cordoned off so no access was allowed by civilian personnel.

Initially, the Air Force released a notice to the public that they retrieved material from a "flying disc". This admittance was then retracted, in classic fashion, lol. A cover story was then fabricated, stating that a military balloon landed on the ranch. This was contradictory to the original information released. Staged photos showing a picture of tin foil were held up for release to the press. In the 1970s, Marcel came forward with the information related to the disinformation and cover-up.

The Roswell incident also had additional key witnesses including Glenn Dennis. This man worked at the funeral home. He met the Air Force while at the base hospital where they had recovered the alien bodies from the crash. The hospital was performing autopsies.

Additional witnesses include nurses stationed at the base hospital. The chief nurse was Lucile Slattery. And nurse Idabelle Wilson was stationed at the base hospital from February 1956 to May 1960. Nurse Eileen Fanton was at the same hospital from December 1946 to September 1947.

In January, on the date of the 7th, in the year 1948, another UFO event made history in America. A captain, Thomas Mantell, was killed while chasing a UFO near Franklin KY. The plane attempted to reach a very high altitude; wherein it was observed that the plane exploded! Mr. Mantell worked for the National Guard. The witness to this event was Glenn Mayes, who lives near Franklin.

The next sighting we will discuss was military. The year was 1956, on August 14. Time 2120 to 2220 military time. Bentwaters England military base, picks up 3 radar tracks. An amber colored object was observed by military personnel through binoculars. The radar indicated speeds approaching 4000 mph. The object maneuvers consisted of stopping in midair, rapid starts and extreme transits. Stan Friedman, a nuclear physicist studied this sighting. Venom aircraft observed the objects travelling below them. The object was circular in shape. Apparently, there was limited cooperation between the British and US military at that time.

The next case is known as the RB-47, a radar and simultaneous visual case. It is considered one of the most important UFO cases in world history.

The incident was examined extensively by Edward Condon, who was the director of the Condon Committee and the University of Colorado UFO project. This is of particular interest because the UFO was tracked on radar as well as emitted radio signals that were detected by the military. The date of the incident was July 17, 1957. The RB-47 jet was undergoing reconnaissance training over the Mississippi, Louisiana and Texas. The Air Force jet uses Electronic Intelligence equipment. This is similar in many respects to air defense radar. It is abbreviated ELINT.

The UFO was measured utilizing a radar beam equivalent of 40 kilowatts. This event occurred for approximately 30 minutes. It was detected in Dallas. The object was located near Ft Worth. The RB 47 aircraft attempted to pursue it but the UFO disappeared from the radar twice. The UFO exhibited radar jamming capabilities. This same capability was also noted in UFO sightings in Canada in 1955, which was investigated by the CIA and released by the Freedom of Information Act in 1989. This was through efforts by investigator Clifford Stone.

The next event was the Brazilian Trindade photos. Trindade is an island 600 miles from Brazil. In October of 1957, the Navy was doing experiments with weather balloons. On January 1, 1958, at 0750 hours, a UFO was observed by the Navy Tow Ship, Triunfo, perferming 90 degree turns and was orange in color, with a spherical shape. At times, the object would houver motionless near the ship. On the 6th of that month, a weather balloon was launched to an altitude of 14, 000 feet. Soon afterwards, military personnel observed a metallic object, crescent shaped and bright white, moving from the southwest to the east.

Another sighting occurred on the 16th of January, 1958. This time it was the south coast of Trindade. An object was observed approaching the island. Witnesses included a civilian, named Almiro Barauna, an expert in underwater photography. The ships dentist, Lt. Homero Ribeiro, also saw the object. The speed of the photographs were 125, with aperture f/8. The object was gray in color and metallic. It had a ring running through the midsection. The craft was surrounded by a greenish colored mist. The photos were sent to the Brazilian Navy for analysis and were identified as authentic. The object was observed near Desejado Peak on the island. The same object was observed by the chief of surgery at Rio, on 230am on the 16th, near the coast of Brazil. Radio transmissions were interrupted by the presence of the UFO. Also, on the ship, electrical malfunctions were observed at the time of the sighting. The electricity went out.

On April 18th, 1961, another unexplained event occurred. This was in Eagle River, in the state of Wisconsin. A round shaped object, like "two soup bowls together", was 30 feet and diameter and 12 feet thick, was observed. It tapered to a thickness of approximately 1 foot at the ends. The witness was Joe Simonton. The object was "brighter than chrome" according to the witness. He also claimed to have seen three occupants of the unusual craft. The occupants were short, approximately 5 feet in height. They were wearing black clothing. This was at his house, 4 miles outside of town. The object made an electrical sound, like that of a generator, he said. When it took off, it went at a 45 degree incline. Dr J. Allen Hynek was the reporter of the incident.

The witness stated that it only took a matter of seconds for the ship to disappear from view. This is a common sight for these craft to exhibit enormous acceleration capabilities.

On December 16, 1966, a sighting occurred in Stillwater MN. It was red, white and green, according to the witnesses. The object was observed near the newly constructed power plant. The object was rotating. The sighting was signed by Gene Lueber. His wife also witnessed it.

In 1964, a saucer was observed landing in Socorro NM. This was dimensioned at 12 feet in diameter. The witness was reputable, a police office named Zamorra. Physical evidence was left in place, indentations in the rock or soil, and some alteration of the existing rock or soil.

The next incident was in Exeter NH. The shape was elliptical. Officer Bertrand described it like an compressed egg. It was concealed by flashing red lights. The lights pulsed in a two second cycle.

It was in September 1965. It happened on the 3rd of the month. This was investigated by the NICAP group of Northbrook IL. Additional sightings were made by several individuals in this area at that time. Mrs. Laroche and Mrs. Pierce also witnessed unusual flying objects. They noticed an objected 12 to 14 feet in diameter at tree top level. They noticed red and white lights emanating from the object. Also, a white beam was directed downward from the craft. They were quite upset by the experience. They described the object as "pancake shaped". The intensity of the lighting increased and decreased. Officer Hoxie confirmed the sighting.

On June 4, 1965, Spacecraft Commander Jim McDivitt had a sighting. This was during a space maneuver. This was during a Gemini 4 flight. The official explanation was that he saw a satellite.

Jacques Vallee, a specialist in data processing, stated that about 8 to 10 percent of sightings are unexplained phenomenon. Mr. Vallee studied about 300 UFO sightings prior to the 20th century.

This included a sighting by the Prince of Wales, on June 11, 1891 between Melbourne and Sidney. The object was described as a fully illuminated ship. Mr. Vallee believes that "amazingly intelligent life beyond the earth can already be discerned" from these sightings.

The next case we will discuss is the Michigan case on March 20 and 21, the year 1966. This event occurred in Dexter. Address of witness 10600 Mckinnie Road. Frank Mannor; at the time of 835pm on 20 of March. Height above ground was approximately 500 feet. The object was blueish green, changing to a brilliant shade of red; and finally yellow. It was reported to the Sheriff. There was a brilliant illumination of the woods around the object. The object took off to the west at a very high velocity. The object was approximately 500 yards from the observer. His son also witnessed it. The shape of the object was flat on the bottom but cone shaped on the top.

Later, at Chelsea, the Police Department was notified of a similar object. The object was the same coloring as the above sighting; it hovered in the woods and took off at a tremendous rate of speed. This was in Webster township; between Scully road and Webster church road. Patrol number 19 was dispatched to investigate. Deputy Fitzpatrick and Deputy McFadden investigated. Additional witnesses included Hunnawill and Officer Bushroe. They observed the object passing near the vehicle. Colonel Miller and Dr. Hynek were involved in reporting and analysis.

The next case, which had a lot of publicity was in the UK. The location was Rendlesham Forest. This was near the RAF Benwaters and Woodbridge military installations. The date was December 27, 1980. Lt. Colonel Charles Holt of the USAF was sent to investigate. The incident occurred between 125am and 400am. A tape recording was made; the noise generated during the sighting was between 600 to 700 Hertz. No bird or insect noises were apparent. There was an actual landing observed of the unidentified object in the woods. It is believed considerably more information was obtained on this sighting by the military but the information was not shared with the general public. The object was observed on the ground for four hours, it hovered briefly above the ground and then took off at an enormous rate of speed. Brenda Butler investigated the sighting after communications with a security officer at the base complex. This investigation lasted three years. Interestingly, the USAF was more forthcoming about the sighting than the British Military. The landing site was adjacent to the Boast Family House. When hovering, the object was six feet off the ground. The dimensions were three meters across and two meters high. Lighting of the object consisted of a pulsing red light on top and blue lights underneath. The animals located near the object on the farm were stirred up into a frenzy during the sighting. This was a significant enough event that the Base Commander went to observe the sighting.

Additional sightings in the State of Utah. In April 2007, at the Henry Mountains in Southern Utah, a large metallic object was observed hovering above the mountain. It was 530 pm in the afternoon. The object was over 100 feet in diameter and elliptical in shape. It hovered above the mountain, which is located in Hanksville UT. The object was observed tilting at a 45 degree angle, and took off, covering 300 miles of airspace in four seconds.

The next Utah sighting was that evening, while the observed was camping in the Henry Mountains just south of highway 24 and west of Hanksville, east of Cainesville. It was 1030 pm at night. A glowing white object was observed gently descending to the southeast side of the mountain. It made no discernible noise. There are no trails or roads on that side of the mountain.

In July 2011, another UFO was sighted in the Captitol Reef National Park area, on the east side of the Park along the dirt road travelling north.

Time of observation was 130am. A camper was awoken by the bright light of an object, 30 feet above the ground surface, totally motionless, along the escarpment of the rock face which borders the valley of the Park. The object remained in a motionless state for approximately 45 minutes before taking off. No noise was noted. The object gave off a blinding white light which illuminated the valley area, dimensions approximately 10 meters across, circular to elliptical in shape.

Dr. Jack Kasher, retired Physicist from University of Nebraska at Omaha, worked for NASA in 1991 as a technical consultant to evaluate UFO sightings, particularly on the Internation Space Station. He was asked to do an analysis of footage of a moving object which approached the earths atmosphere, stopped briefly and reversed course back into outer space. The physics in question were triangulation of known earth reference points in order to determine the G force on the craft during acceleration back out of the earth's atmosphere. Dr. Kasher stated that the object moved from rest to accelerating over 100 G's of force; truly amazing feat of engineering and way beyond human capabilities. The human limit for G force is but a fraction of that, at 46.2 G's, held by Air Force officer John Stapp. The physiology of the inhabitants of such a vehicle, it was concluded had cartilage instead of bones.

Chapter 3 Photos

Picture taken from Navy pilot observing UFO.

Another UFO photo, note relative position to aircraft and landscape background for reference.

Beautiful detailed UFO photograph.

Another UFO photograph, notice trees and telephone pole for references.

Another spectacular UFO photo. Note trees and illumination.

Classic UFO photograph, vintage. Note farm house and telephone pole for references.

Another excellent UFO photograph. Note car for reference.

Extraordinary UFO photo of fleet, Japan.

UFO photo, note mountain for reference.

UFO observed in park, note bench and trees for reference.

UFO from government archives, note rocks and mountain for reference.

UFO above roadway. Note car on road and mountains for references.

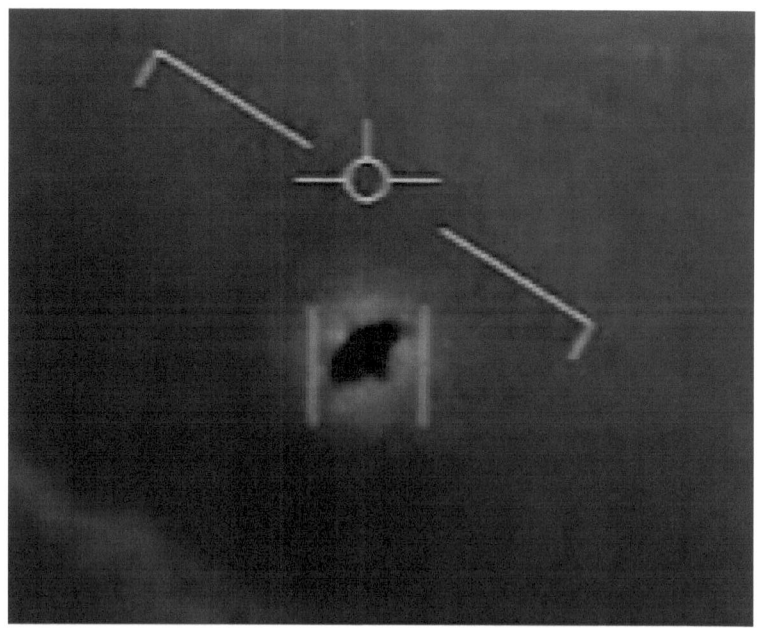

UFO photograph released from Pentagon, Navy files.

Two UFOs above roadway. Remote location. Note trees and mountains for scale.

UFO photo from Archives.

UFO from Pentagon files, inverted shape noted.

UFO photo in wilderness, trees for scale.

Old UFO photograph. Note buildings and trees and truck for scale.

Unidentified lights over water.

UFO sighting, note vehicles and trees for scale.

High altitude sighting.

New Jersey UFO sighting.

Pennsylvanian UFO sighting.

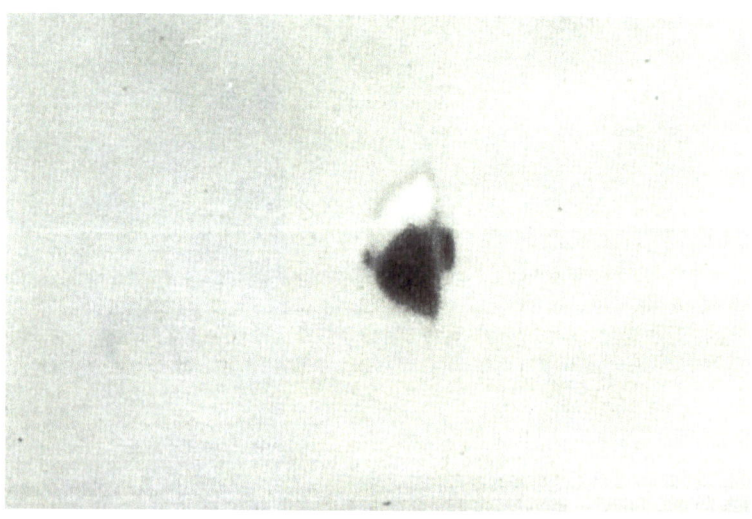

Historical UFO photo.

Historical UFO photo, Orange County.

UFO in New Jersey.

Classic UFO sighting, Santa Barbara CA.

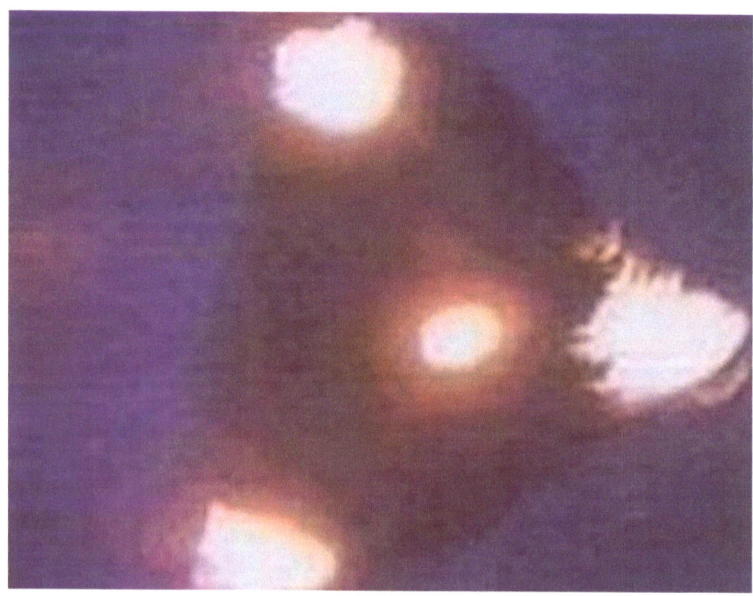

Mysterious triangular shaped UFO.

UFO in Arizona. Triangular shape noted.

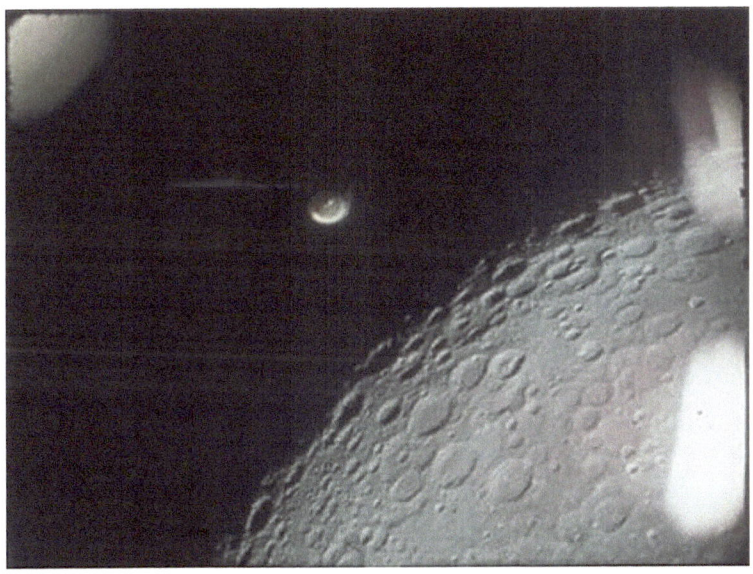

UFO near moon. NASA.

Mysterious unexplained crop circles. Perfectly symmetrical.

Vintage UFO photo.

Extraordinary UFO photo.

UFO photo, India.

UFO photograph in rural location.

Gulf Breeze FL UFO sighting, 30 years ago.

References

Bryan, CDB. Close Encounters of the Fourth Kind.

Clark, Jerome. The UFO encyclopedia.

Corso, Colonel Philip J. The Day After Roswell.

CUFO.org; Center for UFO Studies

Fawcett, Lawrence, et al. The UFO Coverup.

Feindt, Carl. Ufos and Water, Effects of UFOs on Water.

Fuller, John. The Interrupted Journey.

Good, Timothy. Above Top Secret.

Haines, Richard. UFO Phenomenon and the Behavioral Scientist.

Hopkins, Bud. Intruders.

Hyneck, J. Allen. The UFO Experience.

_____. The Hyneck UFO Report.

MUFON; Mutual UFO Network.

Pratt, Bob. The UFO Danger Zone: Terror and Death in Brazil.

Randle, Kevin et al. The Truth About the UFO Crash at Roswell.

Sagan, Carl et al. UFOs a Scientific Debate.

Scully, Frank. Beyond the Flying Saucers.

Walton, Travis. Fire in the Sky.

Webb, Walter. Encounter at Bluff Ledge.

www.ingramcontent.com/pod-product-compliance
Lightning Source LLC
Chambersburg PA
CBHW040330220526
45473CB00009B/2628